WAKE UP, WE HUMANS!

WAKE UP, WE HUMANS!

Read the writing on the wall

PAUL CLIPSON

THE WE HUMANS PROJECT

ISBN: 978-1-919605-60-9 (paperback)
ISBN: 978-1-919605-61-6 (ebook)

Published by
We Humans Press
4 Nathaniel Walk, Tring, HP23 5DQ
United Kingdom
www.wehumansproject.org

THE WE HUMANS PROJECT

To

Will Reed

(1954-2016)

The Survival Sweepstake 2100

Babies born in the year 2020 will turn 80 in the year 2100. Here are the latest odds for them.

		2021	Update the odds yearly
		↓	↓
	FOSSIL RECORD Life continues without humanity which falls before the 2100 running of the race.	3/1 jt fav	
	DODGED A BULLET We 8 Billion humans make it through to a stable future with the same population size as now.	11/1	
	BY A THREAD Five million humans survive in isolated enclaves after multiple pandemics leave us decimated.	16/1	
	ELON'S EMIGRES A colony of survivors on Mars is all that is left of humanity.	66/1	
	DONE AND DUSTED Planet Earth is not fit for habitation by anything – no microbes, no plants, no life. All gone.	66/1	
	HAPPY GO LUCKY Human population healthily in balance with nature. Reduced fertility throughout 21st century means population is steady at 1 Billion.	3/1 jt fav	
	ROSWELL RESCUE Aliens land in New Mexico and solve all our problems for us.	150/1	

Contents

Preface
The story of a dodgy leader

The writing on the wall is an old story that tells of a bad leader who brings a whole country to its knees. Sometimes the story is called "Belshazzar's Feast".

King Belshy was your typical boss – full of himself and always focussed on the bottom line. It was always about money – or in his case – gold. He was so single-minded. Nothing else mattered. The people didn't matter. The land didn't matter. Only wealth mattered: his wealth.

The night of the feast he really went too far. Everyone thought so but no one had the guts to speak up. All those endless toasts. Round after round in praise of the God of Gold. It was embarrassing. Uncomfortable. Crass.

No one noticed the writing on the wall at first. Then we all pretended not to understand it. We all knew it wasn't good news for King Belshy. Old Daniel was the only one who got it. Came right in and told the king straight:

"The writing on the wall – it's for you, King. It says you're not up to the job. Says you're going to get fired any day now. It says your time is almost up."

Amazing how the old man stood there, calm as you like, and laid it all out. We didn't change, of course. Carried on feasting because that is what the king wanted. But Old Daniel was right: by the next day King Belshy was gone forever. And so were the rest of us.

I guess we humans are still enjoying the feast, still drinking toasts to the God of Gold. Still frightened to tell our kings "enough is enough." We don't need Old Daniel, or an old Sir David A, or a young Greta T to read the writing on the wall. We humans need to wake up and read the writing for ourselves. This book is about leadership and about how we follow our leaders. If we don't read the writing on the wall and change how we lead and how we follow then our story is almost done. Like King Belshazzar and his people, we will be gone forever.

1

We humans: An Introduction

1.1 Setting the scene

We humans are in a perilous place. We might be about to die out and take life down with us.

The crisis is urgent and of our own making. We are heating up the planet and we are killing off a vast range of other species that ultimately we need for own survival. The underlying problem is leadership.

Our leaders are operating in the wrong way. They are doing too much: they are too central, too involved and too powerful. The more power they have – the less power we have. We get

1

used to being powerless. We won't solve the climate crisis if we have surrendered our power. Every time a lying, self-serving political leader is elected then we are moving closer to extinction. Cynicism is the understandable response to dreadful leaders but cynicism doesn't make us powerful. It won't help us make the changes that are needed.

Leadership in the world of work is more of a mixed bag. Some bosses really understand the role of a leader and do their best to deliver what is needed. Many of us will have had experiences at the other end of the spectrum: working for a bad boss is depressing and sucks joy and energy out of our lives. Every time we go to work for a bossy, demanding, secretive leader who leaves us feeling bruised – well, that is weakening our own muscles, depleting our power. Those muscles will need to be in peak condition if we are to get out of this environmental mess.

The way out of this scary story isn't to do away with leaders. Leadership has always been one of our superpowers. We have to do leadership very differently to address the challenges that we are facing. Leadership isn't the only thing that will have to change. We humans have an awesome array of superpowers. Leadership is certainly at the top of the list – but it has a twin – followership. We have to be active, challenging and engaged in how we follow – not obedient and resentful. Leadership needs to be more humble and followership needs to be more active if we are to wake up and escape extinction.

Leadership and Followership together are a powerful human adaptation. There is another critical superpower combo in the Human Story as well – it's telling stories and listening to stories. We have this fascinating adaptation: we tell stories, we listen to them and we believe in them.

Leadership skills and followership skills, storytelling and believing in those stories: these are powerful adaptations that allow us to work together as large groups of strangers. We join together using these combined superpowers to achieve epic results. We build cities. We conquer continents. We fly to the Moon. We do so many things with a collective will. This will is rooted in our leadership and followership abilities: that's how we organise ourselves to get stuff done. And where there is leadership and followership, storytelling is always close at hand.

These superpowers have made us a highly successful species but now we are in trouble. Our leadership, our followership and our storytelling are all out of balance. They have been hi-jacked and we need to claim them back. Leaders spinning poisonous stories, followers who are mis-led or despairing – this scenario leaves us in great danger. We are taking a lot of other species down with us. The leadership crisis is dangerous for the whole planet.

This isn't just a recent problem. It's a tendency that began long ago but up to now we have always managed to re-invent leadership to make it fit the challenges of the time. Now the challenges of the time are global. Leadership is still stuck in a mindset of "us-against-them" competition. We need to reinvent leadership, so it has a global perspective: mutual benefits rather than "we-win-you-lose".

Re-inventing leadership goes hand in hand with re-inventing followership. These two human capacities always change in tandem. They need to change again if we are to wake up to our current problems. And they need to change very quickly.

The Worry List

So much worry, so much fear
Are we all doomed? Is the end truly near?
Our carbon fuels lead to soaring heat
And we're all eating too much meat

The honey bee is in decline
And tuna fish – well the fault is mine
We're really bad news for the albatross
And gorillas are going, due to habitat loss

The coral reefs will soon be gone
My worry list goes on and on
What can we do, O Homo unwise
To avoid our own species' demise?

All go vegan? Well, that's a start
End all the methane from when cows fart
And the plastic in fish
Means they're not a safe dish

But eating more beans just won't be enough
The choices we face are extremely tough
Stop burning the oil, the coal and the gas
Stop flying, stop driving and travelling en masse.

Even then – well, it looks pretty bleak
As human numbers are yet to peak.
But we need to retreat, let nature return
Leave vast empty spaces where no carbon can burn.

It's the mighty green forest that keeps us alive
We have to take less, let wilderness thrive
It's pretty simple as long as we're smart.
Is this the end – or is this the start?

1.2 Are humans heading for the exit door?

So where shall we begin our human story? Let's begin at the end … extinction.

Species like ours don't last long. There have been many upright apes with big brains over the last few million years. These species are all gone. We are the only ones left. Some of them have been very similar to us – not just in the same Great Ape family but in the same human genus. The last of the other humans died out quite recently. 30,000 years isn't long ago on an evolutionary timeline. Now we are heading in the same direction as the Neanderthals and the rest: Extinction.

We have been the most successful of the human species: largest population, greatest geographical range, greatest extractor of materials, greatest manipulator of those materials, widest range of adaptations to different environments, widest range of foodstuffs. The CV for *Homo sapiens* is pretty impressive. Slow career at first: couple of hundred thousand years ticking over in hunter-gatherer mode. A lot of travel but no major innovations to make us stand out from our cousins. But we have certainly changed that in the last 10,000 years. A turbo-charged career in every way. No stopping us – unless we stop ourselves. Which is where we are now – our success is bringing us down. We have over-played our strengths and now maybe we are the ones heading for the exit door.

What's it like going extinct? Well, at a personal level it won't feel like extinction. It will just feel like death. A pretty important death given it will be our own personal, once-in-a-lifetime experience of dying. But the thought of all humans disappearing – well that makes a bad day really bad. To add to the gloom, it could be even worse than that. We have kicked off an extinction spasm that is decimating the number of species we share the planet with. These massive collapses in the diversity of the living world have happened before. This one is at least the sixth mass extinction event in the fossil record. Typically, these collapses are followed by an explosion of evolutionary experimentation. New species emerge to take advantage of the new environment. But maybe not this time. There is an outside chance we could take the whole show down with us. If we keep heating up the Earth's protective atmosphere, we may boil it away completely. Mars had a thick atmosphere once. There was flowing water. Maybe even life. But not anymore.

No protective gas skin means no water and no life. Right now, we are shedding our planetary skin. The Earth's atmosphere is heating up because of human activity. Something has to change. Either we change our behaviour, or we humans disappear. If we carry on like this then the Earth ends up looking like Mars. We don't want that; Earth is the blue watery planet, not the red dusty planet. Martian isn't a good look for us.

There is a chance we will survive beyond this century. But there is a chance we are the last humans. There's even an outside chance we are the last act of "Life!" – a show that has gone on for four billion years.

We haven't gone through the door marked "Exit" just yet, but we are heading in that direction.

Neander T Hal
@NeanderH

Really hurts how you guys use Neanderthal to mean
"primitive"
"backward"
"dumb"
At least we didn't chop down the whole forest that we lived
in and then complain it's too hot and there's no shade.
How sapiens is that?!
#WeHumans

1.3 So what needs to change?

We have achieved so much in the blink of an evolutionary eye – a very young and hugely successful species. We got ourselves into this mess using our human skills. Our skills are our only hope of getting out of the mess.

Some people respond to the doom-and-gloom of global warming and the destruction of nature by denying there is a problem.

> *"Global warming is natural. Species die out all the time. Nature can cope."*

But assuming nature will rescue us probably isn't going to work.

Some people respond by being stupidly optimistic –

> *"The scientists will think of something – just you wait and see"*

But passively relying on the brilliance of science isn't likely to work either. There are so many different challenges threatening the basis of life that a single scientific solution isn't going to appear out of nowhere.

Some people respond by being powerless.

> *"Yeah – we're probably doomed but meantime I've got to get on with my busy life. I can't change anything myself"*

Being passive and despairing isn't likely to save the situation – more likely the despair will become a self-fulfilling prophecy.

Some people think we can be better people.

> *"Well, human nature needs to change then –*
> *it's wrong to carry on causing this destruction"*

Human nature isn't going to change.

If we do get out of this alive, it won't be because we become "better people" through some moral revolution that changes human nature. We are what we are. We are a despicable species to cause such damage. We are a wonderful species to achieve so much.

We humans have an awesome ability to transform stuff: forests into fields, minerals into metals, wild into tame and petrochemicals into power and plastic and pretty well anything we fancy. Amazing!

But right now, our transformational skills are warming the air, turning the oceans toxic and disrupting the basic processes that keep the planet habitable for us and our fellow non-human earthlings. Despicable!

We won't evolve our way out of this. The human nature that got us here is what we have to work with. Nothing more – but nothing less.

So are we all doomed?

No – there is a way forward. We may not be able to change our nature but we can change our actions using the abilities we have now. It is human behaviour that gives us the best hope of survival.

Technological fixes may help in the margins but it is human behaviour that drives that innovation.

The way to respond to the crisis isn't through denial or despair or false optimism. The best response is to work with what we have in order to make the big changes that are needed.

What is it in human nature that has made us so spectacularly successful? It is our behavioural adaptations that skill us up to rapidly spread innovations and use them collectively on massive, shared activities.

What is it in human nature that is now driving us to the edge of extinction? – It's the same set of adaptations. We do things on a grand scale and these activities impact the environment. We innovate relentlessly then quickly scale up the innovations. In just a few hundred years we have spread innovations across the planet which have led to global warming and the destruction of nature. We adopt new technologies such as the internal combustion engine, or industrial scale fishing or the chainsaw: these new possibilities spread and quickly have an impact on our atmosphere, our oceans and our forests.

Now we need to quickly spread new ways of organising and new ways of relating to the natural world that repair the damage we have caused. We need new ways of coming together that create a sustainable future for our own species and the many neighbours we rely on in the Animal and Plant Kingdoms.

Innovation, mass collective activity, rapid adoption of new ways of coming together – these social adaptations are not single, specific, narrow behaviours. They are a composite of many elements of human communication skills, human

intelligence and human social organisation. These adaptations are what got us into this mess, but they have also made us successful, and they are the best hope we have for getting out of this crisis.

This book focuses on perhaps the most important adaptation of all: leadership. But leadership would not work without its twin: followership. Followership is an active and highly sophisticated adaptation as well. How we do leadership and how we do followership need to change a lot if we are to avoid extinction. Both adaptations rely heavily on storytelling: the telling of a story by a leader and the shared belief in that story by the followers.

Leadership, followership and stories. It may sound like an odd combination. They are the drivers of our whirlwind human history and they offer us a way out of our current predicament.

2

Working with our Superpowers

2.1 Leadership

Different ways of organising have emerged throughout history– but leadership in some form is a constant.

For over two hundred thousand years we humans have banded together in groups of people much larger than a single family in order to get stuff done: kill a mammoth, build a temple, construct a 10,000-mile wall or fight a war. We organize to get stuff done – it's what we do. All these mighty achievements have involved some kind of leadership. We keep coming up with new ways of organizing. Tribes. Villages. City States. Nations. Empires. These examples are all based on a physical territory.

Sometimes we create ways of organizing that are based on ideas and not just on controlling territory: Religions. Corporations. Aid agencies. Terrorist groups and many more ways of organising that spread beyond physical borders.

We try one way of doing it. Do it that way for a while. A few hundred years maybe. Ditch it for something else. The last 10,000 years or so has seen lots of change in how we group together: Egypt under the Pharaohs; Rome under the Emperors; The Vikings; Benedictine monasteries: all were different ways of organising to get stuff done. All involving some kind of leadership. All involving many people over time and space. Each person did not know everyone else – they were mostly strangers working together on large scale projects. Each system had its own set of rules.

When it's time for a new way of organising human affairs – a change of political system, a change of dominant group, a new set of beliefs – it can get nasty. The disruption of adapting to a new way of coming together is usually massive and often brutal. But we do carry on as a species in our relentless need to get stuff done.

It isn't just the personnel who change when there is a breakdown and a new order emerges. Whole new forms of leadership emerge during the disruption period between one social structure and the next. We know from research into the surviving hunter-gatherer tribes that even in the days before written history, leadership took many forms. When we started writing things down about 5000 years ago, we began to get a history which was very centred on the leader. History has been dominated until recently by accounts of leaders and struggles to change leaders. How leaders behave and how they are viewed keeps changing, but leadership is always present.

The history books never stray far from discussions about leadership:

Who will be the next leader?

What did the previous leader achieve?

How did that leadership change come about?

Our species does things on a grand scale – and the super-power of leadership is key to that ability. We can work together on projects that span the globe and take several lifetimes to complete.

Many other social mammals display leadership behaviours but their selection method is consistent and the role is fixed. We humans have found multiple ways of selecting leaders and have varying expectations of the leaders that we select.

We have been particularly varied in finding leadership behaviours that can generate sustained achievements involving vast numbers of people.

In our ordinary conversations we debate who should be our political leader or the leader of our favourite sports team. Even though we will probably never meet any of the candidates, we debate how they should lead. But we never debate why we have leadership in the first place. Leadership is so obvious to us that we struggle to imagine what it would be like without any leadership at all. Even with small-scale and local activities, leadership behaviours tend to emerge.

Leadership is such a recurring feature of human activity that it does not seem remarkable.

Human leadership behaviours are unique in that they work across time and across vast distance. They work between strangers who may never meet. Most importantly – they keep changing. We repeatedly adjust how we display leadership behaviours. Leadership is a constant that constantly changes in how it is displayed. The problem of the 21st century is that social organisation has changed radically but our leadership behaviours have failed to change at the same speed. Our political and corporate leaders are still operating in the old-fashioned way. We followers become cynical and disenchanted as we have not found a way to express the modern followership that is required to tackle the planetary challenge. Leadership and followership need to change swiftly and radically so we can all take up the challenge of planetary survival.

TAKE ME TO YOUR FOLLOWERS! #1

2.2 Followership

A key without a lock would be of little use. The adaptation that gave us leadership skills would be irrelevant if we were not also adapted to be followers.

We can be fanatically loyal to a leader we have never seen, a leader who will never know our name. This loyalty is not because of their abilities as a leader – it is because of our human abilities as followers. Followership is the twin super-power of leadership. For large parts of our lives, we are all followers. Organizing on a grand scale beyond the limits of personal relationship requires sophisticated followership skills. We are skilled at choosing leaders and remaining loyal to them. We also remain loyal to other followers who share our allegiance. We may or may not benefit from our loyalty but at a population level it enables large groups of strangers to work collectively to achieve a goal.

Another element of our followership skills is mimicry. We are a copycat species. We imitate others in how we look, how we talk and how we think and feel. We look for others to copy. We like to fit in and belong to a group, a gang, a neighbour-hood or a party. Followership skills allow these groupings to emerge. We pick up on social cues and adjust accordingly so that we do not cause offence or reveal our ignorance. Mostly we want to fit in. We do a lot of monitoring of other people's behaviours and adjusting our own to make this happen. These are all key to being successful followers.

Blind obedience is just one kind of followership behaviour. Compliance based on fear or absolute control is another followership behaviour. But these extremes have limits to how long they can sustain large-scale human activity. They are also limited in the quality of what can be achieved.

Sophisticated followership behaviours have emerged that demonstrate tremendous commitment to the leader or leaders and to the other followers. It is not just leadership behaviours that manipulate extra contributions from followers. Followership is as intrinsic a human skill as leadership. Both require a wide range of subtle and complex social skills. Both have appeared in many diverse ways across human history. Followership is a highly active set of behaviours requiring awareness and decision-making.

Combining leadership and followership

One sophisticated trick we have as humans is that our leaders can also be followers: complex projects are organised through this hierarchy of leaders who are also followers. These loyal leader/followers have people who are loyal to them. Again, this is so obvious to us because it is so familiar – but it is a very sophisticated adaptation that has enabled coordinated activity to spread across vast areas. Groups where everyone knows everyone else are quite small. Leaders generally struggle at knowing more than fifty to a hundred followers.

A hierarchy of leadership with followers-who-are-also-leaders fitting into the structure allows humanity to deliver awesome achievements and devastating destruction.

As individuals we can use one set of skills to be the leader of a group, then go into another context and be a follower who is highly alert to the senior leader and their needs. At a micro-level, many changes are visible as we move from the leadership role to the followership role. Our vocabulary changes, patterns of eye contact change, anxiety levels change – our physiology responds differently when we are in the presence of a leader compared to when we are in the leadership position ourselves. These leadership and followership adaptations are intertwined. They are fundamental to our success.

Leadership and followership are also usually implicated when things go wrong. Both have to match the needs, the demands and the challenges of that particular moment. When the leadership and followership behaviours do not adapt to match the challenges of the time then instability follows. The fall of civilizations recorded throughout history often have a natural element such as infectious disease or climate change leading to famine, but there are always leadership and followership issues to amplify the catastrophe.

It's important to emphasise that all of us are capable of leading and of following: it isn't the case that most of us are born to follow and a few are born to lead. That's a nonsense that is popular in certain political quarters but it is not relevant. Leadership emerges in so many different ways and so does followership. Both are key parts of being human.

Today's dominant version of leadership and followership is out of touch with what is needed. Some leaders are shifting to a more participative style. Some are adopting a sustainability agenda. These changes may be heartfelt or they may be cosmetic but the majority of political and corporate leadership is still heavily "old school". Leaders still dominate the narrative. The highly skewed share of an organisation's payroll budget that goes to a few leaders is an indication of their centrality. They may talk "participation" but they still get the lion's share when it comes to pay. The emphasis is still on attracting and keeping the right leaders. "Who is the right person to lead?" is the wrong question.

We need to find better ways of leadership that take away the attention and the power and the dependency on a few individual leaders. Leadership is a necessary and efficient process. It should not be an opportunity for maximising the centrality and the wealth of a few. The problems of our age affect all of us. All of us need to come together to tackle them. Powerful followership is key to that process. This participation will not emerge when power is held tightly by a narrow leadership caste.

The Winning Ticket in the Carbon Lottery

Coal, gas n' oil – a real big win
Turn those engines, make things spin.
We hit the jackpot – all that power
Work took a week – now it takes an hour

We Won! We Won! Let's Spend! Spend! Spend!
A carbon rush right till the end
We scooped the Lotto, untold wealth
But carbon riches are bad for health.

We're supercharged on endless fuel
But longer term we're nature's fool
Like all big winners we'll crash in a while
Stinking rich but short of a smile

It's hard to adjust to instant riches
Friends are lost and everyone bitches
Now Mother Earth is very cross
And in the end she's still the boss.

2.3 Storytelling

The anthropologists got it wrong when they named our species Homo sapiens ('wise man'). In any case it's an arrogant and big-headed thing to say, wisdom being one of our least evident features. In reality, we are Pan narrans, the storytelling chimpanzee.

Terry Pratchett, *The Globe*

We are not the only social ape but we are the only one that tells stories. Not only do we tell stories but we listen to them, we believe in them and we act on them. They have been a potent part of our dramatic success and of our most ugly failures. When we see other people believing in their stories, we think they are misguided or deluded. When we believe in a story ourselves … well, it's because we are right. Once we believe in a story, we are very loyal to it.

Throughout our history we have told stories. Throughout our history we have listened to stories and remembered them. Stories aren't a diversion at the end of a busy day. Stories are central to how we come together in large groups to get stuff done. A successful story needs a skilled storyteller but it needs skilled listeners as well – a story without listeners is no story at all.

Stories help us work together, to respond to change and to quickly spread new ideas. Storytelling, story-listening and believing in stories are part of being human. People who believe in the same story have a deep connection that fosters loyalty and collective action.

Believing in the story is crucial to how we behave

It isn't the story itself which is crucial – it is the shared belief in the story. Right now, we humans are in trouble. It doesn't look like our story will end well. But a lot of stories are scary – which makes them more gripping. The scarier the better. A good story needs intrigue and danger, heroes and villains. And then a happy ending. And if happily-ever-after is not possible because the hero dies saving the day, well at least we want the story to end in hope for those that do survive. We always need hope so that more stories can follow.

There is plenty of danger and excitement in the unfolding drama of humanity in peril. The climax is getting closer. Soon we all could perish. Wiped out as global temperatures soar, ecosystems collapse and species around us tumble into extinction. It's difficult to imagine a happy ending given the current predicament. As with all exciting yarns, at the darkest point it can feel like no hope is left.

Unless you don't believe in that story. Belief is very import-ant. Without belief, stories don't hold us. We lose interest. It's not about whether the facts are right or wrong. It's about engagement. Factually we know there aren't short, stocky creatures with hairy feet who set off from Hobbiton carrying a magic ring. But we believe in the story – or at least suspend our disbelief – for over a thousand pages or the length of three feature films because what happens to Frodo and Sam matters to us. We are inside their story. But if we don't believe, if we don't get swept up – then we are left wondering what all the fuss was about. "The Lord of The Rings" doesn't do it for everyone.

"The Story of the Climate Emergency" doesn't do it for everyone either. They don't get swept up by the flow of the story so the statistics leave them ... well ... cold! People can look at graphs of CO_2 emissions and global temperature without seeing it as part of a plotline leading to catastrophe. But it is hard for those of us who have entered into the climate story to understand how others are not moved by the grave dangers that lie ahead. Impending doom is not believable if you are not inside the story. Extreme storms that cause catastrophes are seen as individual events that are not related to other catastrophes. They are not trends. The deadly bushfires in Australia, the relentless forest fires in the Pacific Northwest of the American continent, the melting of the Antarctic ice, the increase in the strength of hurricanes – all these are key parts of the plotline for some, but for others these events are not building to a crescendo of danger. They don't need a heroic response because they are just disconnected events that are unfortunate but independent. One person's stories are another person's coincidences. These contrarian voices place themselves in a different set of stories that paint the climate activists as the real threat.

The climate story that one believes in determines how one views the data – instead of the data determining how one assesses the risk. Stories have such a powerful hold on our way of seeing the world that they even influence how we interpret facts about the weather! Stories influence which facts we seek out. Stories influence who we believe and what we believe. Whom we listen to and whom we choose to ignore.

Shared stories are part of that adaptation of working together on complex projects such as carving out an empire or constructing a high-speed train line. Stories have been at the heart of these massive endeavours. Facts are judged on their accuracy and don't galvanise people to behave differently. But stories are judged on the feelings they trigger. If the story hits the right feelings, then it is a powerful glue to join people. If the story-glue is strong enough then boring facts won't break people apart.

Stories have been central to our success as a species of collective achievement. The underlying skills behind this human superpower of storytelling are complex. The imaginative power of the storyteller, their ability to memorise, to communicate emotion and character and plot are all part of the puzzle. The story-listeners have reciprocal skills that combine to create the human phenomenon of mass shared belief. But the powerful magic of story really comes to the fore when we see how it combines with the other superpowers of leadership and followership.

Neander T Hal
@NeanderH

I never thought you guys would come to much when you wandered into our valley. You looked lost. And so puny! And it looks like I was right, I guess!
#WeHumans

Leadership, Followership and Storytelling - a winning combination

3.1 Leaders tell us stories - and we followers listen and believe

This story shall the good man teach his son;
And Crispin Crispian shall ne'er go by,
From this day to the ending of the world,
But we in it shall be rememberèd
We few, we happy few, we band of brothers.

Shakespeare's Henry V talking to his men
the night before the Battle of Agincourt

When we add storytelling to the leadership-followership combination then we end up with a hugely powerful adaptation
that separates *Homo sapiens* from the rest.

We are an unusual species that can work together without
knowing each other and while we are spread out over distance
and across time. When required, we can rapidly change how
we organise: new leaders and new ways of leading, new stories and new loyalties emerge.

Leadership and followership are both sophisticated adaptations that use many skills. They are two sides of the same coin;
complementary skill sets that make up the *yang* and the *yin* of
how we accomplish large-scale activities. A similar relationship
exists between the storyteller and their listeners. Sophisticated
skills are used to bring a story to life – to make it believable.
Other species have leadership and followership skills – but they
tend to be used in a local way. When you add stories to the
mix then collective activity can involve strangers who may not
even be alive at the same time. Stonehenge was constructed
across a thousand years. The builders of medieval cathedrals
would be rushed to finish their work in less than a century. The
possibilities for this kind of extended collective activity appear
infinite. And that's the danger, as we will see later.

As followers, we choose which leaders to follow partly on the
basis of the stories they tell. Stories have always been a key
element of the pattern that links leaders to their followers. We
bind ourselves not only to our leader but to their stories and
to other followers who believe in the same story. It's natural
to us: we like to hang out with people who believe in the same
stories as ourselves. We follow leaders who tell stories that
excite us and that resonate with our view of the world.

That is why many of our stories are about leaders. We may not love our leaders but we certainly have a thing about them. We write about them, talk about them, focus on them. We adore them or bad-mouth them. And leaders use stories to do their job. The Human Story tells of leaders achieving great things and tells the stories told by those leaders. Changes and disruptions in how we organise ourselves generate new stories that reflect and reinforce the new ways of living and working together. These new stories are often entwined around the new leaders, their origins and their achievements.

Clever leaders use storytelling to create bonds between people. When we get swept up in a story, we can be fiercely loyal to the leader who is telling the story. We also become loyal to the other followers who share belief in the same story.

We can be loyal to the point of death in pursuing an imagined ideal, a visionary goal painted by the leader. Their story is often highly emotive but vague, exciting but untestable. Loyalty flows in all directions: to the story and to the leader and to the other followers who believe in the same story.

Shakespeare – the master storyteller – made a career writing stories about leaders. He even embeds stories within his stories. He famously tells the story of the Battle of Agincourt in which the young King Henry V tells a story to his troops about how they will tell a story many years from now as they look back and remember their bravery tomorrow on St Crispin's Day at the Battle of Agincourt. Stories within stories within stories. Past and future all mashed up together. We can get mesmerized by the multi-layered way that leaders use stories. Sometimes they hark back to previous glorious times. Sometimes they evoke images of a glorious future. Past and future get blended into one emotionally charged present. Stories allow us to play with time and twist it. Our human minds are adept at responding to emotional cues that come from a leader. Logic and fact and current reality give way to a mental state which is more nebulous and open to suggestion and – most importantly – it is shared. Leaders use stories to instil the important belief that we are all in this together. Loyalty, obedience, collective action may all flow from the emotional experience of the shared story. Leaders aren't all being manipulative when they use this technique. Certainly, we can find examples of such behaviour where the leader is cynically telling lies to trick people into a course of action. But there are many examples of leaders who are swept up in the telling of the story so that they believe the truth of their own words. The argument here is not one of truth vs lies. Truth has a different status in stories.

Stories are especially powerful when the storyteller is also a leader. By entering into the story, we get a strong sense of belonging and participation. We know that once we enter into the story our reality is altered and we make different judgements about facts and data. Leaders are able to exploit that shift from our logical minds into our story-minds.

The night before Agincourt things didn't look good for the English. They were outnumbered three to one. They were facing annihilation. Henry mesmerized his followers through story and the next day they stuck together and fought bravely and won. Maybe the story helped Henry as well.

Storytellers use the story to draw their listeners into a shared experience. Leaders draw their followers into an imagined experience of achieving the goal. If we believe in the goal – the story – then the bonds between leader and follower become extremely strong.

The role of "Leader" isn't the same as the role of "Hero" in the story. But there is a lot of overlap. Our shared human reality is highly receptive to heroes and to leaders. We give them special status. Leaders gain from having one foot in the world of story. If the leader introduces a heroic element to the group's goal, then the followers also get bathed in a heroic light.

Leaders who talk in heroic tones create multi-layered loyalties to the leader, to the story and to the other followers of the leader and the story. Leadership and storytelling have so much in common – they occupy adjacent plots of land in our mental landscapes.

We humans consistently look at a world where we expect to see these particular plots. Stories are fundamental to our way of seeing the world. Leadership and storytelling are massively over-lapping human traits that have been key to our whirlwind success as a species.

3.2 Leaders and Heroes in today's world

No more heroes any more
Whatever happened to
All of the heroes?

The Stranglers

Today we don't set our stories in enchanted forests and medieval castles. Twenty first century battles are set in corporate castles and political forests. That is where the heroes are triumphant and the villains are vanquished. Right now, it seems the villains are having an easy time of it.

There is a whole industry for teaching leadership yet we seem to be short of good leaders. There are leadership colleges and leadership consultants and leadership programmes all showing the way to the high status role of being the leader. One prestigious business institution is promising to teach people how to be "authentic" leaders. Surely their students have already failed if they have to be taught how to be authentic? Leadership may have been there since the start; today leadership is still valued and seen as a highly modern set of behaviours. Business successes and political successes are attributed to the leader's endeavours. In Medieval times the health of the nation and the health of the king were seen as one and the same. The mystique of the leader still lingers around 21st Century board rooms and it is, more often than not, still a king, a man, who is on the corporate throne.

In the business world we don't give the same status to people who are "managers". It is a less exciting activity. Leadership is connected to inspiration and passion and vision. Leaders put on heroic clothes in a way that a mere manager would never be allowed to do. In politics it is the same: many of the people who do the actual work are dismissively labelled "administrators". It is the leaders who face the exciting challenges, who slay the dragons and capture the castle.

Business leaders and politicians are keen to be seen with military heroes and sporting heroes. There is a conveyor belt taking Olympic gold medallists into corporations to teach and inspire future leaders. The inference is that rubbing shoulders with heroes is enough to confer some heroism on to corporate leaders. Leaders often use the language of battle or sport to exhort their team to succeed. They ask that people "go the extra mile" in winning against the odds. Obstacles are turned into opportunities. Achievements are epic. Failure is not an option. People are ignored when they are steady or efficient; they stand out when they are driven and uncompromising. Success is heralded with fanfares and honours – even when the quest is simply to achieve Quarter 3's ambitious sales target for the baked goods category.

Sometimes we don't love our leaders – we despise them. They can generate great passion and excitement or they can provoke fear and indignation. But the key is that leaders typically produce a strong emotional response. We react and respond to leaders. They invite us into their story of how the country or the workplace will be under their leadership. That's the thing about leadership: we react to it strongly.

These strong emotional responses to leadership are familiar to anyone who has worked in a large corporation for a long time or followed the merry-go-round of party politics. In recent US politics there have been extreme emotional swings as Bush gave way to Obama who gave way to Trump who gave way (or refused to give way!) to Biden. There have been similar dramas in the UK as the leadership story moved through an unlikely cast of Cameron, Corbyn, May and Johnson. Each stole the headlines in successive years. We have strong emotions when it comes to our leaders and how they behave. These strong responses are not new. They have probably been there since the start of humanity. We may not love our leaders – but we are seldom neutral about them.

Successful leaders are aware of this emotional response and are able to harness it to achieve their goals. We are not naïve as a species. We are often aware what our leaders are up to. But we are still biddable – we can be persuaded to do what is asked. Words from a leader are potent, regardless of how true the words are.

Let's look again at how we physically respond when we meet a leader. In a meeting at work with a senior person typically we are more alert and sensitive to the leader's mood than we are to other people in the meeting. We anticipate the meeting and are hopeful that it will go well. We are keen to please and hyper-aware of any displeasure that the leader displays. We talk about that meeting afterwards.

On a larger scale if we ever meet – or even just see – a national political leader, we remember it forever. If we meet the CEO at work this can produce trepidation, elation or anxiety. Our physiology responds to leaders. Even if we don't want it to happen we can blush or stammer or sweat more when we are talking in front of a leader. Our eye contact unconsciously stays with the leader. The internal conversation in our heads is commenting on how well or how badly we are performing in front of the leader. Everything is different when the big boss is in the room.

Likewise, the stories that fire our imagination have a space for a leader. Our biology has a reaction to a leader. These followership traits have been with us for a long time. They make leadership work.

Twenty first century leaders are using new channels to talk to their followers but they are still wrapping themselves in the cloak of the hero. They may invite us to listen to their stories on a Zoom call but we are still falling under an old spell.

PUBLISHED IN LONDON, NEW YORK, TOKYO AND READ IN ALL FINANCIAL CENTRES STILL ABOVE WATER

FINALITY TIMES

ANY DAY NOW   FOR THE PEOPLE WHO DECIDE   10 SATOSHI WORLDWIDE

 " How dare you blame us!" — THE OLD GUARD FIGHTS BACK, PAGE 21

Boardroom struggle at BigOil inc. — Dividends v. Descendants

In happier times

Another setback for Musk Mars rocket — no electric vehicle charging points at destination

Story on Page 8

Bitcoin mining now uses equivalent electricity of whole of USA — and rising

Reflection: Does greater good include non-humans?

Continued inside

3.3 Why we waste time on make-believe

Maybe words really are that powerful. Or maybe that's just a story we tell ourselves to imagine power into existence. Everything you know is a story.

George The Poet

Stories are powerful. They shape how we view our past and they shape how we move forward to create our future. Why? We can play at evolutionary psychology and guess why stories are so central to the human experience but there isn't much hard evidence. Ancient human skulls can give us some indication of the structure of our ancestors' brains but the stories inside those brains are not in the fossil record. We do have some evidence that stories have been around for a long time. The oral tradition has preserved stories from the ancient past. Our species has been hunter-gatherers for over 90% of our history. Recent lifestyle changes in the last 10,000 years won't have changed our DNA that much so we can reasonably assert that we have hunter-gatherer brains. And the remaining hunter-gatherer societies still have strong storytelling traditions. It is a reasonable deduction therefore that if a particular behaviour – storytelling – has been there since the earliest days of our species and has persisted right through the invention of printing and through the Golden Age of Hollywood and into the Age of Netflix and binge-watching then it isn't just a minor and incidental habit. It must have an evolutionary significance. Why use all that mental energy and time on something so unproductive?

Over the millennia of human evolution "wasting time" on stories must have worked for us. It must have given us some evolutionary edge. What makes stories so powerful?

- Stories are not fact-based though they may have some connection to the facts.
- Stories are not just the territory of children and irrational adults.
- They are not a lesser kind of thinking but a different kind of thinking.
- Stories are rooted in belief not in logic.
- Stories hold people through emotion not through data.

Many hundreds of thousands of modern-day Hindus, Jews, Christians and Muslims retell stories which can be traced back in some cases to the Iron Age or even earlier. We humans have got to where we have got to because stories work.

Sharing a story makes us feel special. We are on the inside. We are connected to the storyteller and to the other listeners and to the story.

When a large group of people get swept up by a political or religious movement, they become immune to other voices. Observers on the outside express disbelief at the implausibility of the story they are all listening to. Crowds gather to hear the storyteller, preacher or politician. These fervent followers bathe in the excitement of shared belief and shared loyalty. However, the skills of the storyteller do not need to include truth-telling. They need to create a sense of shared understanding and shared destiny that separates their followers from those outside. This is often easier to achieve if reality is abandoned completely. The more fantastical the narrative, the more special it is to be an insider.

This human capacity to be swept up by a shared story goes beyond the charismatic power of the storyteller. A good story may not be rational, but it still somehow 'makes sense'. On the outside we may ask why would people be so gullible? How can people believe such nonsense? But those questions miss the point. It is not that the shared story is false – it is that it is shared. Humans get tremendous psychological reward from belonging. It feels good to be on the inside, to be special, to have the secret knowledge.

Stories are highly contagious. They are highly emotionally charged. They create strong bonds between the people inside the story. They create strong barriers against the people out-side the story. And this happens with incredible speed at key times of uncertainty. Our "gullibility" is a human capacity to swiftly organise in the face of adversity. Unity has typically been more useful than accuracy in the face of external threats.

Let's speculate that in our ancestors' time maybe changes in the food supply made a long journey the best way to survive. The tribe did not scientifically assess the climate change that could be impacting the availability of protein from seasonal hunting. They just noticed the herds of reindeer were smaller and arriving later each year. A shared story of an angry god who needs appeasing may quickly get people to undertake a perilous journey to new hunting grounds. Success on the journey will come from everyone sticking together. If each family had made their own choice and acted independently then the collective strength would have been diminished.

So stories bind people together, they hold people together when times get tough and they provide a vehicle for new ideas to quickly spread among strangers.

We won't find ways forward by lamenting how deluded people can be when they fall for dangerous stories. The remedies proposed in this book won't work if we think stories can be replaced with facts. Beliefs cannot be corrected with reason. Facts come into it – but we humans follow our hearts. The leader woos us by telling stories. Our hearts have a soft spot for leaders who tell stories and for other people who follow the same leader and listen to the same story.

We Leaders, We Followers, We Storytellers and We Story Believers have this powerful, dangerous way of coming together to achieve massive things. We can't eradicate this combined set of adaptations. Why would we, when it has served us well? We need to use it more intelligently if we are to collectively turn back from the brink of extinction.

Don't judge me

I don't give a toss about the albatross
Starving polar bears – who really cares?
I know bleaching coral isn't moral
But anyway ... It wasn't me.

Sea levels rising? – not really surprising.
Ozone depleting? Ice caps melting?
All such a shame but I'm not to blame
Because you know – It wasn't me.

I care so much about things I can touch.
My lovely wife and our way of life.
Our kids – all three – and their favourite tree.
They really matter because they are a part of me.

But I worry at night, wake up in a fright
About things that I read and get on my feed
This environment stuff is hard to rebuff
I'm getting scared – just between you and me.

The kids are clever – they say "it's now or never"
That the Planet is dying and soon we'll be frying.
Then my kids and my wife would have no kind of life.
Protecting them – that's what's worrying me.

It's so hard to say what will save the day.
I'll fight if I must – but who do I trust?
Show me how and I'll make a vow
You know – you can count on me.

4

The dark power
of stories

4.1 Us vs Them - how stories recruit
people into opposing groups

Children are the vessels into
which adults pour their poison.

Salman Rushdie, *Midnight's Children*

We humans adapt swiftly when circumstances change. It is hard to prove it – but let's keep playing with the idea that stories are an essential part of human societies across all of our history because they are key to navigating change.

Stories set expectations that difficulties will be encountered and heroically overcome. Stories are central to our species because they are a powerful mechanism we use to come together and stay together when times get tough. Collective action doesn't happen if every individual makes their own independent decision. Under pressure, or in a rush, we are more likely to abandon logical thought and default to our "story-settings" and go with what we have been told by the people that we value.

We look to the leader. Believing in their story is crucial to how we behave. It seems to be a short cut in our individual decision-making process. The leader looks back, knowing that keeping everyone together is critical.

In a crisis, the group has this adaptation to bind together. People on the outside of that tight matrix are seen as wrong or as threats or as unimportant. We have seen how groups form and bond around stories. Members of other groups weave their own different stories. Listening between the two camps goes down; barriers go up.

All too easily we can get divided into opposing camps. The story that we believe in determines how we see other people who are not living the same story. We are swift to label people outside of our story as "The Enemy." Having a shared enemy is part of the bond between people inside a shared story.

There are so many examples of this division into opposing story-camps. Awful genocides are documented throughout our history. Looking only at the last 100 years there are many examples of dark and menacing stories being spread by one group about another. Germans were fed stories about the mis-deeds of the Jews in the 1930's. Hostile stories about the alleged behaviour of Muslims circulated among the Hindus during the partition of India. Malicious rumours fuelled the tensions between the different ethnic groups in Rwanda before the genocide of the Tutsis.

In each case, groups who had co-existed peaceably ended up with the less powerful group being the victims of terrible violence. It is a complex human question that is hard to understand: why do people who have been decent and moral in their dealings with their neighbours then collectively turn on them with such awful consequences? Malevolent stories spread by malevolent leaders are part of the explanation. Stories can bind people to their own group even at the expense of individual morality.

There is something embarrassingly simplistic about this tendency. In the 21st Century surely our thinking is more sophisticated than "Goodies v Baddies"? Unfortunately, it still permeates how groups interact with each other. A key division of our age was mentioned earlier: Climate Change Activists vs Climate Change Deniers. One group sees the other as false prophets of eco-doom; they are foolish and gullible and probably acting as unwitting agents of a sinister foreign enemy. But to those on the activist side of the climate change story the reaction is very different. They see climate change deniers as the enemy. Their refusal to see global warming as a threat is seen as morally wrong and is contributing to the danger that we are all in. In America a significant part of the population still disputes the existence of the climate emergency. "The Story of the Climate Emergency" is becoming a fault line in American society. Believe global warming is an existential threat and you belong to one camp. Believe it is a dangerous liberal hoax and you belong to another camp.

It is not just an American problem. Each of us arranges our reality to fit in with the stories we listen to…. which then become the stories that we tell and then they become the stories we believe in. We are inside these stories; the stories belong to us and we belong to the stories. The story one believes in is tribal in its consequences: the other tribe is seen as alien. One only listens to storytellers from within one's own tribe. So each side becomes ever more entrenched in its own story. And neither side is very successful in recruiting people from the other tribe. Changing one's mind may be easy but changing one's story feels like betraying one's tribe.

Stories don't usually bind the whole species together. They are effective at creating groups of humans that are opposed to other groups of humans. They separate the human world into insiders and outsiders, allies and enemies. This becomes a real problem when the story needs to embrace the whole species – not just one group of people against another group of people.

Great storytellers achieve a universal reach that embraces all humanity. They may write about different tribes and warring nations but they help us to see the threads that connect us all. These profound insights are sadly lacking in most populist leaders whose vision does not extend beyond the borders of their own group.

Large-scale and co-ordinated group action has left us dominating the planet. But the group that has acted together has not been humanity in its entirety. Dividing the species into sub-groups such as a tribe or a nation or a religion has worked well for organising collective action up to now. This division into opposing sub-groups does not equip us to tackle climate change or the destruction of the natural world. We look foolish as we consume our own life support system because stories are dividing us. We live in a time when the threats are global. Many of these stories are misleading us at a time when we have to make a collective shift in how we live our lives.

Neander T Hal
@NeanderH

Homo sapiens has 4% Neanderthal genes – just hope that's enough. You sapiens are going to need all the help you can get from outside as you don't help each other very much!
#WeHumans

4.2 Don't bother me with the facts

Storytellers play with time, they imagine happy endings, they paint pictures of brave heroes – but they don't deal with facts. Facts are deadly serious and don't lend themselves to being played with.

There is a human tendency to coalesce around highly charged stories. Leaders who just talk policy and logic do not generate the same fervent following. They are seen as too cerebral, too dry.

"Make America Great Again" or "Take Back Control" are 21st Century slogans that became tribal in the loyalty and the resistance they engendered. These became powerful stories woven around particular leaders and their ability to deliver a certain ending.

The promised outcome from this kind of story is not a set of policies. They seldom have details of how we will get there or what the precise outcome will be. The future – or the restoration of some previous mythical greatness – is left vague but it has a strong emotional content. There is a sense of longing for something heroic. The aspirational mood leaves people feeling validated and recognized. If we enter into the leader's story we will be embraced by this happy future. We will be a part of the story with its shared perils. The baddies will be vanquished. We will live happily ever after.

When one is on the outside of these potent political movements it is easy to be condescending and dismissive of such falsehoods. But if we are to move away from the cliff-edge of extinction we need to understand the attraction of simplistic stories that pull people in the wrong direction. Leaders who are factual and earnest and unemotional do not have the same impact. They get labelled as "policy wonks" or "lacking charisma". It seems sensible that a leader would talk through the details of their plans but it doesn't have the same traction as an emotionally charged story. Criticizing a story by checking the facts doesn't make people doubtful about the story. It reinforces their belief and places the doubter, the questioner, on the outside.

Stories aren't enough on their own. Leadership and followership have other capacities requiring detail and planning and hard logic. Leaders who stay only in the storyteller space are dangerous. The mass emotion and unity whipped up by a leader without a plan can quickly turn ugly. These leaders specialize in fostering hostility against outsiders. They polish resentments rather than offer any clear way forward. Leaders who can infect a crowd with their stories but have no clear plan, accomplish very little but often leave a legacy of bitter divisiveness.

There has been little subtlety in the politics of the first decades of this century. Populist politicians have been using stories to line their followers up in opposition to other groups. Nationalism is still hugely attractive even though our real problems have little to do with national borders. The rise of the nationalist strong men has been a theme of our time. Their stories have common elements:

*"**We** are the true citizens. **They** are a threat. Together we must stand firm against them."*

Populist stories are weak on the detail of complicated problems. So are many of our current politicians.

Stories don't have to deal with messy detail. They tend to work best with a high emotional content and not too many boring facts. They are a useful tool for a leader. Stories unite followers to all move in the same direction. But stories are not a good tool for analysing the detail. They tend to be wonderfully vague and gloss over the tricky bits that don't quite fit.

The story-telling adaptation works to generate collective action not to generate shared wisdom. A story can lead a whole group up the garden path.

Anyone in the group who shouts: "Hey, Are you sure we are going in the right direction?" can be branded a trouble-maker or a worry-wart.

Adapting to climate change will have a lot of tricky bits to deal with. We need leaders who can galvanise collective action **and** who can handle detail and complexity.

Storytelling is a powerful leadership tool. Like all powerful tools it can be dangerous if it is used for the wrong job.

MEANWHILE, AT THE PARIS CLIMATE TALKS ...

4.3 Nature is a threat in many stories

If you want your children to be intelligent, read them fairy tales. If you want them to be more intelligent, read them more fairy tales.

Attributed to Albert Einstein (but maybe that is just a story)

Stories happen in nature. Childhood storybooks largely depict an outdoor world to children who are predominantly growing up indoors. Nature in these stories can be fearful.

"Follow the path through the woods
Till you come safely to the other side.
Don't stop to rest.
Don't drink water from any of the streams
And whatever you do...
Don't leave the path!"

The old stories focus on the challenge that the hero has to overcome. "Don't leave the path" is the instruction. But of course, they would be pretty dull stories if the hero didn't leave the path. Nothing would go wrong – but nothing would be learned. No discoveries would be made. The hero would not need to be cunning or brave or be rescued or do something that helps them escape the dangers of the forest. Forests are always wild; they are filled with hidden threats. When you stray off the path then things get out of control. And when they get out of control, the story takes the hero into danger. Staying on the path is the human way. The controlled space. When you stray off the path you enter the non-human world. Things get scary. Wildness takes over.

We humans have had an uneasy relationship with nature since we settled down and began living in villages. We selected which wild animals we would tame and which we would eliminate. We decided which wild plants are crops and which are weeds. We have been very successful at domestication. We feel safer when wildlife is domesticated. As human populations have grown, the wild spaces have been reduced.

It's scary leaving the path and going into the wildwood – so we have tamed the wood. Reduced it, converted it into ships and housing and fuel. Wildness has been replaced with ordered land where we choose what will grow. Wheat for grinding or palms for palm oil or pasture to grow meat or uniform rows of fast-growing trees to harvest into paper.

We have won – nature has lost. Our numbers have gone up and the numbers of so many other species have gone down. We have replaced the wild spaces with human spaces. But now the story is taking an interesting twist. We need the wild places. That's where the carbon we produce can get recycled. That's where the microbes and insects and plants that we depend on can thrive. Safety is in the wild places.

We easily default to a combative stance in relating to the natural world. We battle against storms; we conquer mountains and we fight the floods.

Nature is part of "Other" – that group which is not us. We need to adjust our mental landscape to see ourselves as within nature, not above it or outside of it.

Stories are the interior decorators of our mental landscape so the stories we tell the kids and the stories we enjoy as adults need to encourage new emotions:

We have all left the path.
Now everyone is in danger.
What an exciting time!
Stay wakeful. Be ready. Be brave!
Accept that we are part of nature
so not everything can be controlled.

We need these new stories of hope in which humanity is placed within the natural world rather than seeing ourselves as outsiders.

4.4 Stories told around a different campfire

Stories are a very strong glue – but if you get super-glue on your fingers or in your eyes then you have a real problem.

The massive increase in communications in the Age of Information has amplified the strengths and weaknesses of our storytelling human adaptation. Stories are told around virtual campfires. Listeners do not sit in a circle with the darkness behind them. They log on and peer at a screen.

The new methods of communication have produced little common ground between opposing groups, no listening or mutual respect. Societies are divided on petty issues that will become irrelevant if there is no longer life on Earth.

The internet and the smartphone have speeded up communication and favoured those voices that are succinct, uncompromising and partisan. Followers can filter what they want to hear and who they want to hear it from. Story-tribes are more sharply defined than ever. Social media has exaggerated the high emotionality of stories. We are more separated into hostile groupings, less driven by clear thinking and more dismissive of the detail. At a time when we need to act with unity, we have invented new ways of being divided.

These innovative technologies provide new ways of influencing through sharing stories and new ways of forming group loyalties. The early uses of social media may not be encouraging – but it is still early days. Daily newspapers began in the 19th

Century. The mass communications of the 20th Century were radio and television. Social media is still a teenager. Facebook launched in 2004. The first iPhone hit the market in 2007, a year after the first Tweet. Instagram launched in 2010. This rapidly changing communication channel brings opportunity even if its early history is disappointingly divisive.

There are reasons to be hopeful:

- The speed of adoption has been very rapid.
- Social media behaviours are constantly being modified and refined.
- The reach has been global – it is a new set of human communication behaviours that is emerging in all ages and across most societies.

If social media undermines knowledge and facts and contributes to spreading the dark stories of "Us against Them" and "Let's Conquer Nature" then this communication channel will be a powerful and dangerous tool. If it redefines storytelling, leadership and followership it has the potential to be an amazing human phenomenon that brings us all together to do what needs to be done.

Social Media:
Useful Tool or Scrolling for Doom

As we scroll and swipe through real life
And Digital Me is all you can see
So much is vacant, so much is trite
But it speeds things up in this survival fight.

While some of it's ugly, nasty and cruel
Just see it as neutral – another human tool
That brings us together to get stuff done
Stand and Fight! – or Turn and Run!

Stories get spread with super-charged speed
This gives us a chance in our time of need
Joined in a web of like-minded strangers
Standing together, facing terrible dangers.

Storytellers around the digital fire
Now cast their spell using internet power
We are recruited to tribes we never can see
But I know you – you're on board with me.

Some pitch "Superior Us" against "Enemy They"
with dangerous tales leading us all astray.
Using the web darkly to divide or deny:
"We are not part of nature" – that old, old lie.

Let's share the truth: "We are all in one tribe
In a web of connection that keeps us alive."
Let's refresh and refresh this old wisdom story
So the narrative ends not in death but in glory.

5

Crazy People – Crazy Times

5.1 We don't really understand Time.

"Max sailed back over a year
And in and out of weeks
And through a day
And into the night of his very own room
Where he found his supper waiting for him
And it was still hot"

Where the Wild Things Are. Maurice Sendak

Crazy people have always thought the world was about to end. The world was supposed to end in the year 66 of the Common Era. And again in 365 and 400 and 482 (27th May to be precise) then in the year 500. Then a gap before apocalyptic predictions resumed for year 793, year 847, year 992 … and so on.

These doomsday expectations are a recurring theme that affect individuals and even gain some traction with larger groups. Most people ignore them and get on with their lives, assuming tomorrow will be much like yesterday. Worrying about the long term future isn't a mass obsession.

In fact, we humans don't always take time seriously. We postpone doing important things. We deny we are growing older. We can live our lives pretending we are immortal. We assume our way of living will largely continue as it is now. Then a plague descends on us, a flood sweeps through, invaders come and a civilisation falls. Something always comes along to mess up our assumption that we live in a stable and predictable world – but until the disruption arrives, we prefer to pretend all is well.

We humans are not well equipped for facing up to extinction. Doomsday prophets are a bit of an embarrassment. "Longer Term" means next year. "Strategic" is a popular word with leaders but it seldom reaches more than three years into the future.

Our story-telling brains don't help. In stories, time gets stretched and altered.

- A year-and-a-day.
- The Thousand and One Nights.

- Once upon a time.
- Then. Now. The Future.

In Storytime we don't have to worry that it will be too late. In stories there is always just enough time for the hero to arrive and save the day.

But in this current drama, the one about the end of life as we know it, we are running out of time. Don't let the reassuring comfort of stories give us false hope. The danger isn't imminent – it has arrived. The crisis isn't approaching. It's here. A few years further along this path and we could reach tipping points in our natural systems. Irrecoverable loss of habitats. Arctic warming leading to ice melting leading to more warming in an unstoppable positive feedback loop.

Looking forward, we don't have the luxury of time.

Looking backward, it has taken a long time to get to this point of near-extinction– but most of the really dangerous trends are very recent.

Time has an important part to play in understanding the crisis we are facing yet we are not good at understanding time. History isn't everyone's favourite subject – but you do have to look back sometimes in order to understand the future. Looking back at all those Domesday predictions of "The End of The World" doesn't teach us much about our own period. But if we look back across the whole of human history then the message is screaming at us.

WARNING! ALERT! DANGER!

To understand this message, let's squeeze the whole of the history of our planet into two pages. Then the lesson becomes very clear that we live in exceptional times.

Unless we are real nerds, many of us tend to glaze over when it comes to dates. First human species? Was that 2 million years ago – or 200,000 years ago? (answer: probably *Homo habilis* more than 2 million years ago) Most of us forget whether our particular kind of humans – *Homo sapiens* – have been around for 500,000 years or 100,000 years. (Somewhere in between is the answer)

Let's go right back and begin at the beginning – just in case you missed any of the previous episodes...

Previously in the Human Story...

◉ **13.7 BILLION YEARS AGO**

The Big Bang

Cosmological Time

◉ **BILLIONS OF YEARS PASSED...**

Our Sun, This Earth

Astronomical Time

◉ **SEVERAL MORE BILLION YEARS PASSED...**

Planet Earth changes in slow time. Life appears. Change measured across billions of years.

Geological Time

◉ **MILLIONS OF YEARS PASSED...**

Species change when circumstances change.

It takes two lifetimes to have any evolutionary consequences. It takes many lifetimes for those changes to accumulate into a pattern. If you are a bacterium then a generation can last 20 minutes. Complex mammals like humans have generation times measured in the decades. Change accumulates in our genetic record but it is not visible across one lifespan. Hominids as a family emerged more than five million years ago. *Homo sapiens* just emerged as a species a couple of hundred thousand years ago or so. Since then we spread across the continents of the world and appearances changed. Surface changes in response to pressure from sunlight or moisture in the air. But nothing major. Still the same species under the skin.

Evolutionary Time

◉ **TENS OF THOUSANDS OF YEARS PASSED...**

Our lifestyles didn't change that much either. We adapted to different climates and food sources in different parts of the world and developed different social structures around that. But essentially, we followed the meat supply of wild animals and gathered wild plants. Steady lifestyles across many generations of Hunter-Gatherers with occasional bursts of new things to deal with. Knowledge accumulated and was shared. Traditions developed that valued stability and continuity.

Anthropological Time

◉ **THOUSANDS OF YEARS PASSED...**

Amazing stuff started to happen about ten thousand years ago. We started choosing to live around animals that we could domesticate. Then we started to choose which plants to domesticate. We started to stay put for longer periods. Villages rather than camps. Pastures rather than hunting. Growing crops rather than gathering. These changes didn't all happen at once. They didn't happen only in one place. But over the next few thousand years we made big changes in the human way of life. Most of us stopped living as nomadic hunters and adopted different variations of an agricultural life. New social structures developed. New knowledge. New traditions. Massive changes across several continents across only five thousand years.

Archaeological Time

HUNDREDS OF YEARS PASSED...

Now things really start to speed up. Larger scale societies developed in multiple settings. Across the last four millennia we have seen complex and sophisticated civilizations emerge: China, Egypt, Persia, the Ancient Greeks, the Romans, the Incas and the Aztecs and others. They all lasted for a period then slowly or dramatically faded away. A few hundred years seems to be the timescale as some humans organised in very large societies and experimented with urban living. Innovations develop and spread; social structures and ways of life don't last as long as they used to.

Historical Time

TENS OF YEARS PASSED...

At each stage of the Human Story we have been successful at extracting more resources from the environment. Food, wood, metal, fibres. We have taken natural materials and converted them for our use. Energy has been key to that extraction and conversion process. Three hundred years ago that process went through massive innovation at the start of the Industrial Revolution. Humans learned to accelerate the extraction and processing of raw materials to create goods and release energy. We have been riding the consequences of that innovative wave ever since. The speed of those changes is unprecedented. Our species has made many changes in lifestyle, in social structure and in physical environment. These changes have now spread across the world. Stability has given way to waves of further change at a speed that we have not previously encountered.

It took us 2,000 years to get us from the wheel to the printing press. 300 years more to get to the steam engine. Another 100 years to get to the light bulb. 50 years on we invented computers. Then in quick succession, within one person's lifetime, we invented space rockets, desktops, the internet, laptops and smart phones.

The product development cycle time? Modern Time? No time at all?

Each cycle of change is faster than the previous one. More change to assimilate with less time for us to learn how to work together in a new way that fits the new reality.

This squeezed summary of all of time tells us one vital thing: We humans are really in a hurry. Species don't normally change their way of life this quickly. Our speed of change has been very fast for the last 10,000 years and it is accelerating.

We humans have done rather well. We had a way of life that didn't change much for a long time. Then there was the first big change – settling down into an agricultural, village-based life. There were about 5 million humans at that point. The start of the next massive change – the Industrial Revolution – saw a planet populated by 500 million people. Only a few centuries later and there are nearly eight billion of us. Over half of us now live in cities. We became a majority urban species in about 2008.

There is an argument that we have been very successful at managing change. Populations have increased. Our lifespans have increased. For many, our living conditions have improved. We adapted well to change when we became farmers then adapted on steroids to the more recent changes as we moved through the Industrial Revolution and then changed again into the Information Age.

That's the optimism – that we are hugely competent at creating and handling change. We have seized each opportunity to adapt and our species has thrived in the new circumstances. The adaptation time has shrunk from millennia to decades. The changes were initially incremental – now they are fundamental. And still we march on.

This has spilled over to affect the other species that we depend on – and those species are not change-junkies like us. They need longer to adapt to new environments. We haven't changed our genetic code – we've just changed how we organise. Other species use the much slower change process of evolution. That's why they are getting caught out as we chop down the forests and increase the temperature of the seas and the air.

And if they disappear, then we disappear. That could happen in no time at all.

We have arrived at a critical point. Maybe all generations think their time is the most significant, the most urgent or the most unprecedented, the most challenging or the most difficult. Our time now can claim all of those descriptions. But we haven't had time to adapt to the recent massive changes that got us to the present day. It's very exhilarating racing downhill, feet going really fast, trying not to stumble, only keeping upright by moving ever faster. But if the slope doesn't start to level out then the speed of the descent will be too much and it will end in tears. We have created so much change by our ability to extract and release carbon. We have impacted fundamental biological systems by our expansion into the wilderness spaces. Our success has been at the expense of the rest of the living world.

It's a real thrill racing downhill until the moment the velocity is too great. That's where we are now in the Human Story.

Do we get grazed knees and elbows, or do we wipe out completely?

Creating new ways of living on the planet isn't just about changing our food supply. Recycling more won't do it. Buying green products won't do it. The change requires fundamental shifts in our entire social structure, in the way that we organise ourselves. We have to do different things in a different way. It takes time for these new ways of being in the world to be tested and modified and rejected and then for the successes to flourish and spread.

The well-funded optimism coming from the Climate Change Deniers camp is spreading a dangerously different story. In the end we have to choose what we believe in. For many of us the time is now. The crisis has come. We have to respond accordingly.

Neander T Hal
@NeanderH

Saw this painted an old wall
"As you are – so were we As we are – you will be"
Believe me – extinction sucks.
#WeHumans

5.2 If the End of the World is nigh-so what?

It's the end of the world as we know it
It's the end of the world as we know it
It's the end of the world as we know it
and I feel fine

R.E.M.

Emotions are an important part of being human. If human nature is going to rescue us from this environmental jeopardy, then our emotions will be a part of it. They haven't really kicked in yet.

Our emotions are not engaged by "parts of CO_2 per million" or "bee populations have declined by 76% in 20 years". The scientists who are describing the current eco-catastrophe need help. They are describing a reality – not telling a story. Their description is boring with its graphs and its measurements.

Our misguided belief that we are not part of nature has dulled our response to the destruction of the natural world. The separation between ourselves and nature has been a painkiller, reducing our sensitivity to the destruction of all around us. We know all this stuff about environmental devastation but we still carry on doing what we do. It's not that we are stupid – we wouldn't have achieved so much if we were – it's because we are not sufficiently connected to nature at an emotional level.

We humans have lots of emotions. Stories are a collective way of sharing and channelling those emotions. But our traditional stories have depicted nature as a threat. Now our emotional responses are confused and blunted.

We can deny the mess we are in. Many people do.

Feeling scared is pretty sensible but probably won't help on its own.

Feeling guilty probably won't help either.

Feeling angry is understandable – but who do we blame?

Feeling excited might be more useful but only the most extreme thrill-seeker would feel excited in the face of such danger.

Blended together, all these emotions enrich the story and give it a sense of urgency. The only dangerous emotion is denial. Refusing to feel our feelings never ends well. It would be better to get really scared and hide behind the sofa. Or get really angry and point the finger at the baddies. Or get really guilty and point the finger at ourselves.

We are responsible for both catastrophes: the climate crisis and the destruction of the natural world through human expansion. The combined effect of the two crises can be over-whelming. Both crises are a threat to our continued existence. And in the best traditions of the blockbuster disaster movie the consequences of both threats are coming to a head simul-taneously. The narrative arcs coincide: the climate emergency and the extinction spasm are peaking right now.

One emotion that needs to flood through us is sadness. The losses already are tremendous. Entire ecosystems already destroyed. Wonderful species lost forever. Teeming oceans reduced to marine deserts. Grief is the overwhelming response.

With so many strong emotions it is tempting just to switch off. Nail-biting drama can be like that. For most of us we have other things to worry about as well as the planetary catastrophe. We worry about our small concerns. We know it is foolish to worry about how we will pay back student loans or pay off the mortgage or save for retirement when all the while the world is spinning toward annihilation. But we still do it. Our personal worry lists differ but we share the global worry list. Sometimes it is like listening to two phone calls through one set of earbuds: the left ear is hearing a caller demand attention to the items on the personal worry list and the right ear is hearing a different caller being insistent we attend to species going extinct and the collapse of entire ecosystems. The two conversations can only be understood if we alternate our attention or rip out one earbud. Another way to stay sane is to take out both earbuds and block up our ears. But the conversations will soon start again. We avoid the global issues to attend to the personal concerns, kidding ourselves the big problems can be dealt with later or by more powerful people. But there is no time for delay. We are past that point. The time is now. There is no one else who is going to fix the problems. It's down to us. So we end up back in the swirl of emotion and feel overwhelmed.

Stifling our emotions stifles ourselves and takes away our ability to take action. It's the end of the world as we know it … And it isn't fine. We need to feel that it isn't fine. All our emotions need to be felt, to be given their chance to breathe. This emotional expression is happening more and more around the world. This should give us reason to hope. There is a famous story where all the troubles of the world escaped from Pandora's Box. The only thing left at the bottom of the box was hope. Hope is a very powerful emotion. We need plenty of hope right now.

6

Hope for the future

6.1 Superpowers at the ready

Will we get written out of the script? Or will our story-telling brains find a way to escape from peril?

We are realising that our success is turning into failure.

The Human Story is reaching a climax. We are in grave peril.

Our political leaders, the leaders of our corporations, the leaders of our institutions are collectively leading us over the cliff. We humans need to make some massive changes.

Quickly. We need to do more than choose better leaders. We need to make a major adaptation in what we expect from our leaders and what it means to be a follower. We have made large shifts like this before. We have repeatedly found new ways of leading and of following. Stories have been at the heart of these rapid and major changes.

Leaders without followers end up going nowhere. We all have our part to play if we are to escape an unhappy ending.

Stories got us to the amazing, perilous place we are in today. If we are going to escape these current dangers then we will need new stories and new storytellers. And remember – stories without listeners aren't really stories.

We have taken over the world. We have grown our population from 1 billion to nearly 8 billion in just 200 years. We have used our superpowers to extract so much material from the planet that our man-made stuff weighs more than the weight of all the other species put together.

Our leaders got us into this mess. We let them do it. We are recognising this. Belief in our leaders in many countries of the world is declining, particularly in the democracies of Europe and North America. People are more energised by the leader they are against rather than the leader they are for. Enthusiasm for the leaders of political parties has been replaced with cynicism. Some people are still buying in to their leader's story but not in the numbers needed to create a critical mass to support major change. Leaders with minority support may hold sway before they are replaced with the contrarian view of a different minority. There is no sustained and overarching leadership and followership united by a common story. This unity over time has been the hallmark of major historical change. At present this pattern is disrupted. Followers are creating demands but our political leaders are failing to respond.

Not all political leaders are self-serving control freaks – but there are enough of them to feed the caricature. Most politicians are simply not giving voice to the reality of our age. We do not need shallow fools lacking a moral compass who are focussed on winning the top job for themselves. Those politicians are stuck in a previous age.

We followers need to give voice to the suffering of our planet so that our political leaders can hear. Every protest at the felling of an ancient tree or the closure of a neighbourhood library or a protest against a plan to build luxury homes rather than affordable housing – all these voices are the expression of community. These followers are telling our leaders their job is to protect the Common Wealth of Planet Earth.

Many of our current political leaders prefer to only deal in two dimensions – for example they will argue for reduced environmental regulations in order to increase the freedom of business to compete. They simplify issues to "either-or" arguments – then choose which side of the argument to promote and which argument to dismiss. In doing so they move from being two-dimensional to being one-dimensional. Most issues of our time are three-dimensional (at least!) and we need leaders who can manage this complexity. They must not talk in simplistic slogans and mantras. They must give voice to the contradictory pressures and multiple claims that are the mark of our age. Leaders who only march to the tune of the free market would lead us on a shortcut to doom. Uncontrolled market forces deliver outcomes of environmental destruction and extreme inequality.

On the other hand, we know that the imposed and absolute equality of communism crushes the human spirit and stifles the emergence of solutions to complex problems. But we currently have dangerous levels of inequality between countries, between citizens in the same country and between workers in the same enterprise. This inequality undermines the battle for the environment which will only be won by all of us having a share in the victory.

We are a smart species – we can juggle more than two things at once. We need smart leaders who are comfortable with complexity and don't reduce it to a false simplicity of "This is good and this is bad". It is our job as followers to co-create the leadership we need for our time: a leadership which builds community.

We can find examples of this modern leadership approach but seldom do these gifted leaders emerge in the world of national politics.

We need to develop leadership and followership that is suitable for the crisis we have created for ourselves. It's also a crisis we have created for all the other species who are our fellow citizens of Planet Earth. They could get by fine without us. They have their own superpowers – but we would be dead and gone without them.

We have to bring our superpowers into play not only to save humanity but to save the atmosphere, the oceans, the forests and the soil. We think we are the top of the food chain – that makes us vulnerable if the systems that feed us collapse. We have to come together and use our leadership skills, use our abilities to follow in an organised way and our abilities to use stories as a social glue and a mechanism for quickly spreading new ideas. All these human skills will be needed in order to save the non-human species and systems that we need for our own survival.

Our expertise at handling rapid, major change has galvanised human progress so far. We have been experts at making fast changes happen – but not for the good of all. Sub-groups of humanity have benefitted from disruptive change, often at the expense of the majority. The increased productivity that has arrived as we move to a digital economy has only increased the wealth of an elite few. This massive change in how we conduct business has barely altered living standards for most of humanity: the growth of the internet and smartphones and all the related changes have concentrated wealth in the hands of less than 1% of humanity. We are less experienced at delivering fast, major changes for the good of our planet overall rather than for the good of small groups of human beings.

We have the skills but we need to use them in a different way to bring about the changes that are needed. But we are not just watching this story. We are in it. We all have a part to play in the drama.

Better Leadership – what does that look like to me?
Better Followership – how do I participate in that?
Better Stories – how do I make those happen?

Those are the questions we each need to address. This final section of the book points to some of the work we need to do in these areas.

FRATERNITY

#1 JOBS BOARD FOR THE FUTURE

Opportunities for community-based leadership roles

Qualifications: Must have
- Low need for centrality.
- Able to articulate complex and inter-connected problems in an understandable way.
- Ability to co-create solutions through listening and patience.
- Optimistic storyteller who repeatedly demonstrates that we are all in this together.

Useful additional expertise:
- Environmental literacy
- A sense of urgency

Rewards:
- A meaningful life
- Valued as a community member
- Excellent future opportunities

Applications

Apply yourself now through every decision you make – large and small – about your own future.

APPLY   SAVE

6.2 New Leadership and Followership - changing together

Leading and following is a relationship and like all relationships, both parties are changed.

We won't sort out the future just by choosing better leaders. It is much more important to sort out "Leadership" rather than just worrying about getting the right leader. We know that "Followership" and its counterpart "Leadership" are the yin-yang of how we humans organise ourselves. Leaders seldom completely change the minds or the views of their followers. It is a relationship and like all relationships both parties are changed. We understand how an invisible magnetic field works. We are starting to understand how a gravitational field works. There is also a field between leaders and followers. An intuitive leader is alert to that field. They give voice to the fears and hopes that their followers hold deep inside – but have not expressed. Followers are not passive in this relationship: they are constantly giving cues about what they want, what they will tolerate, what they will resist.

As we look at the twin crises of our time – climate change and the extinction spasm – we cannot naively hope that a saviour-leader is going to come who will lead us to planetary salvation. The whole leadership-followership field has to change for us to make the massive changes within the short timeframe before we become extinct.

That's all a bit abstract. Let's break it down:

It's not just about "The Leader"

Leaders always disappoint in the end. It is easier for us followers to have someone who will do it for us – and whom we can blame when it doesn't work out. But that is naïve. We have learned a lot in our brief history as the dominant species on the planet. Leadership is central to getting large-scale change to happen – but leaders don't invent the change on their own and they don't work on their own to make it happen.

Followers aren't fans – they are active participants

A new kind of followership is emerging, ready to step up and take on the challenges we are facing. It is less deferential than in earlier times. It is made up of multiple voices – not just a few look-alikes of the leader. Followers in the past may have been too obedient or have been too passive. Modern followership is speaking what needs to be said without hope of impressing and without fear of the consequences of delivering difficult messages.

The Leadership–Followership field is massively diverse.

There is a hopeful trend emerging – just at the right time. Leadership roles in the industrial era have largely been the preserve of white, straight men from a narrow educational background. There have been repeated challenges but this small sub-group of humanity has held on to its domination. The current way this clique clings to power is by agreeing things need to change – but then carrying on as before. The cries for change have been driven by the injustice of it all. The most recent demands for fairness have emerged under the banners of the Me Too movement, the LGBTQ+ movement and the Black Lives Matter movement.

But now it is not just about the morality of greater representation. A radical new energy is needed to change the Leadership-Followership field. Followers speaking out and speaking up – ordinary people – have brought great energy to these movements for change. Diverse voices have to be heard, diverse leaders have to be in position, so that we humans can do what needs to be done. The narrow, old order simply won't be able to get us where we need to get to. We need diversity because it is the right thing to do – and we need it because it is an essential part of fixing the planetary problem. Leadership rests on a credibility which is undermined if all the leaders look, sound, think and act in the same way.

6.3 Corporations pretend to be communities but we need real communities

The battle cry of the French Revolution had three demands. All three are important.

"Liberté! Egalité! Fraternité!"

Too much focus on individual liberty gave us the excesses of Capitalism. Too much focus on equality gave us the failures of Communism. We must march under the flag of Fraternité.

We can translate Fraternité in many ways. It is more than Brotherhood and Sisterhood. It is Community. Shared endeavour. Mutual trust and responsibility. Caring. The things we hold in common.

Communities have a particular kind of leadership. A patchwork of different players contribute to the overall guidance of the community. Unlike in a corporation, there is no single hierarchical structure organised in a vertical way leading up to the boss-of-all-bosses. There is no singular, all-important goal. Corporations respond to multiple demands but their ultimate rationale is financial: they are legally required to maximise the return on shareholder investment. Communities have responsibilities and aspirations but they do not have an over-arching, specific goal that trumps everything else.

A community in a diverse inner city, for example, may be seeking a number of improvements to local facilities. A group of self-employed artists may form a pressure group to get an old building converted into studios available for rent. The local health practitioners may be working on joining up the health and social care for the elderly living in the community. They will be working with the unpaid local councillors who have responsibility for social care. Leaders from the local mosque and church groups will be involved. Another group of parents with young kids may be seeking funds for a dedicated playgroup facility, and so on. Leadership emerges and moves around. Goals are varied and sometimes in conflict – local council funds cannot support all the projects. This is a community. Leadership and followership are shared and fluid. The balance of priorities shifts from one part of the community to another.

It is easy to make corporations the villains of our story. The reality is far more complex: there are some corporations who are environmental vandals, others that are struggling to do the right thing while staying cost-competitive so that they do not lose customers.

We can all wish for simple explanations of climate change and environmental degradation that do not require us to do anything ourselves. It is very tempting to blame the crises of our age on the Evil Empire of Big Corp inc.

If only it were that simple. Large, multi-national, publicly traded corporations are a very mixed bunch from the villainous to the trying-very-hard. Corporations control so many elements of modern society. They are economically powerful. They have a strong influence on political leaders. They may affect the air and water quality around their factories. They control their global supply chains in highly demanding ways, the working life of their employees can be stressful and they can make decisions independent of the needs of the communities they operate within.

Those corporations that are lagging behind on environmental issues need to step up and put more effort into real change rather than wasting money on green public relations exercises. But the problems are far more deep-rooted than whether Company A has a robust sustainability policy or Company B has a route to zero carbon emissions.

There is a fundamental human need for community that has persisted throughout the history of our species. Corporations are pretending to be communities but corporations are only temporary structures – just another example of how large numbers of people come together to get stuff done. Corporations will change or fall apart. They have borrowed the clothes of community but differ in some significant ways. One example among many is the so-called "Wal-Mart family". This particular corporation made $15billion profit in 2020 but opposes a raise in the minimum wage that would provide many thousands of their "family members" with an escape from poverty.

Corporations are different from communities in vital ways:

- **Corporations choose who belongs and who doesn't.** When times get hard, large companies do not hesitate to reduce the size of the workforce. Downsizing is a recurring corporate behaviour. Likewise, when someone doesn't fit in to the corporation they are kicked out. In many corporations it is risky to ask difficult questions or challenge the way of working. Real communities pull together when times get hard; they don't reduce the population. Real communities embrace dissenting voices rather than banishing them.

- **Corporations relocate to different communities when it suits them.** All too many of our corporate leaders are prepared to move their operations to a different community if they are burdened with too many demands. If the community says "stop polluting" they will manufacture elsewhere. If the community says: "give employees minimum rights and protections" they will build the new office somewhere else.

- **Corporations minimise their community contributions.** Successful workplaces need healthy and well educated employees who can travel easily to work from their safe and secure homes. Communities ask business to contribute to their employees' education, healthcare, transportation system and so on through taxation. Most corporations employ skilled accountants to ensure the tax bill is as low as possible. Any community that asks for too much in tax is at risk of losing the jobs the corporation provides. Community comes low down the list of corporate priorities because there is always another community to move to.

- **Corporations are political players who lobby against community responsibilities.** Large companies are adept at appearing to make positive changes while lobbying governments to minimise changes to the status quo. Poisoning the water supply of a neighbourhood, for example, is not very community-minded. 40 million Americans drink contaminated tap water but many companies still oppose new environmental regulation. Poisoning the atmosphere with carbon emissions is the same behaviour on a larger scale. Corporations often locate in a community because of what they can get – such as tax breaks, minimal controls on emissions or low levels of legal protection for employees. Progress will happen when corporations see themselves as long term community members who lead change rather than block it.

Corporations have a single ultimate goal. They have a clear and linear power structure. A community has multiple goals and outcomes. Communities react to protect the community. A corporation reacts to protect profits. Corporations have a vertical structure which may or may not encourage participation. A community is a participative, horizontal network which, on occasion, needs vertical structures to deliver a particular outcome. Corporate leaders have learnt the language of community but they are still working within the constraints of a system that exists to maximise profit.

The role of leaders inside corporations is very challenging; some corporate bosses try to be community-focussed in how they lead employees but the underlying organisational purpose wins out. Not all bosses are greedy fat cats. (Though unfortunately there are still plenty of those around). Many corporate leaders are trying to balance the impossible demands of their customers, their shareholders, their employees and their suppliers.

Large, for-profit corporations often have great sophistication in their leadership ranks but they are not overflowing with leaders who are rising to the challenges of our times.

The public sector offers more hope. Leaders running large operations in the public sector have to manage complexity and contradiction on a daily basis while binding together underfunded but critical services. These leaders may not seek the limelight but they are often called on to demonstrate higher order leadership skills.

Another place to look for positive signs is at small, privately owned companies free from the greedy demands of shareholder returns. These companies are usually at the mercy of the founding family. This can produce tyranny in some cases; in others it can lead to innovative leadership that experiments with employee relations and community building beyond the factory gates. The tradition of Cadbury's Chocolate and other Quaker-owned businesses were setting the standard of active community membership back in the 19th Century. There have been many more examples in the subsequent 150 years. Sadly, these innovative approaches swiftly disappear when privately held companies are swept into corporate ownership.

Leadership where community matters is not a dreamy ideal. It exists now. We are not looking for an imagined set of ideal circumstances. We need to find the pockets of excellence and broadcast their achievements. Some will be in large systems – public sector or not-for-profit or private sector – and some will be in small scale enterprises. These success stories demonstrate a different relationship between the leaders and the followers. Power is shared in the Leadership-Followership field. These systems already exist and have been proven to work for all stakeholders. There are many places the corporate sector could borrow ideas from if it wishes to truly build community. Some organisations are actively pursuing positive change. Some are very skilled at pretending to change while keeping the underlying power dynamic exactly the same.

In the next section we look at some mainstream ideas for change in the Leadership-Followership field – and some of the tricks that avoid real change.

Reasons to be cheerful

We all have to come up with our own list of reasons to be cheerful or else the environmental crisis can get really depressing. Then we lose our ability to act. "Reasons To Be Cheerful: Part 3." was the original list – Ian Dury and The Blockheads. They had a long list. This list is much shorter. It changes from week to week.

Reasons to be Cheerful: Part 4

♫ [CHORUS]
Every little thing helps, every little thing helps...

♫ [VERSE ONE]
Solar Power, Cars by The Hour,
Organic Flour and ... Zero Emissions
Wind and Wave, Hail Sir Dave,
Eco- Raves and ... Better Decisions

♫ [VERSE TWO]
Carbon reporting, No wildflower mowing,
Green movement growing and ... Save The Whales
Protest, Protect, The Eden Project,
End the Neglect and ... Nature Trails

♫ [VERSE THREE]
Green New Deal, Joe Biden's Real
Plant- based meals and ... NGO's
Serious Greta, Costa Rica
Fishing quotas and ... on it goes ...

6.4 The workplace as classroom

If my experience demonstrates that my voice does not matter then I will learn to be powerless.

If I am told that my voice matters – but then my voice is ignored – I learn to be cynical.

If experience tells me that I can make a difference, then I can develop the skills to be powerful.

The world needs active and challenging citizens who can participate in bringing about change. Leadership and followership skills are key requirements for this kind of participation. The workplace is the major place where adults learn leadership and followership skills – it is the training ground, the daily reminder of what it means to lead and to follow. We need workplaces that develop sophisticated followership – not places that give a recurring experience of resentful obedience. We need workplaces that develop leaders who can handle uncertainty, multiple demands and the need for urgent change. Too many workplaces teach leaders to be single-focussed and driven by short-term financial results.

There are other places – other environments – where adults can learn new skills. But for most of us when we move into the outside world beyond our families, the workplace is the dominant experience. This is where we come together with other humans to get stuff done. This is where we learn leadership and followership. This is where we learn about power.

Progress toward a more active followership in the workplace is not progressing fast enough as we seek to reverse the environmental degradation that is the challenge of our age. Large corporations occupy a position of enormous power. The inevitable changes which are coming will disrupt this corporate power. Large companies are adept at appearing to make positive changes while lobbying governments to delay making changes and to minimise any associated costs. Corporate employers offer a work environment which is often disingenuous: internal communications experts coach leaders in the language of inclusivity while in reality power is held tightly by a select few. This gives employees of many corporations an experience of powerlessness. They have autonomy and awareness in their lives until they go to work when they have to suspend their disbelief. They know corporations are damaging the planet. They also know they need a job and being too outspoken on political or environmental issues jeopardises their job security.

- **Corporations teach obedient followership.** The workplace is a major classroom for teaching followership. We need an active, inquiring and challenging set of followership behaviours to address the challenges facing our species. Many corporations encourage false feedback: "Speak out and speak up" is the apparent message. The reality is often more punitive for those who do challenge upwards. Real communities need assertive participants who are not fearful of the consequences of speaking up.

Employees know they have virtually no chance of influencing the major decisions so they are left disoriented: they have a growing awareness of what is going on in the world and a growing powerlessness to make changes. The workplace skills them in helplessness at a time when all of us need to be discovering our power.

There are many mechanisms for radical changes in the world of work that would make employees true participants. These are well known. They have been tried and tested but few corporations will take them on-board. They prefer occasional symbolic pageants to the hard work of true change.

Mechanisms that make a difference include:

- **Every employee owning a piece of the company**: Shared ownership changes the conversation about what is important. When employees become co-owners then distant shareholders encounter the reality of the organisation that they own.

- **Shared rewards.** When everybody gets a piece of the pie then there is some truth to the inclusive message that leaders like to promote. "We are all in this together" has more credibility when everyone gets a profit share – not just the lucky few at the top.

- **Worker representation.** There are multiple mechanisms to ensure all voices are heard – works councils; worker directors on the Board; Supervisory Boards; trade unions. The goal is that all voices are valued and the decisions are smarter as a consequence.

- **Greater autonomy at a small team level** so that people have some agency in how they do their work. Feeling powerless is a familiar and dangerous response to the environmental crisis. We need to give people expertise at being in control of their own part of the bigger system.

- **Eliminate the corrosive effect of gross inequities in reward**. In the UK it takes top bosses just 3 days to earn more than the annual salary of their mid-point average employee. This rewards profit-generation by the few rather than true wealth creation by the many. It becomes a poisonous and divisive element of corporate life that deflects from the collective action we need to take to tackle real challenges rather than achieve profit targets.

Multiple voices. Wider participation. Better outcomes. It's a virtuous circle.

Unfortunately, leaders are learning the skills of pretending to listen. Mechanisms that make little difference include:

- Specialist advisors and consultants setting up employee opinion surveys
- "Town Hall" consultation exercises
- "Road shows" and all the dark arts of what is called "Employee engagement"
- Performance reviews where feedback is supposed to go upwards and downwards but in reality are about the more powerful voice winning.

The goal of these false listening exercises is not to actually find out useful ideas and information. These exercises only exist to give the employees a sense of being listened to. People aren't stupid – they know when a boss is ticking the box marked "consult with employees" rather than when the employee is being asked to participate in a genuine attempt at co-creating solutions.

It is always obvious when the leader is doing something **to** the employees rather than **with** the employees. When leaders plan these window-dressing exercises the labels go something like this:

"engaging with the workforce"

"consulting with them"

"holding a listening exercise"

And so on. All these leadership behaviours are looking for a particular outcome. They are driven by a desire by leaders to win favour rather than to truly have participative inquiries that lead to shared outcomes. People know this, so their participation is dutiful. They keep their best selves under wraps because it is safer. Such exercises are an inadequate response to the challenges of our age.

These stylised attempts to create a sense of employee participation disguise the underlying reality of a traditional high-control, profit-driven hierarchy. When employees come to work in these stifling systems, they have to leave so much of themselves behind including:

- their environmental awareness,
- their desire for social justice
- an expectation they will be treated truthfully and fairly
- their sense of active participation in the world

We are all in a crisis here. All of us. If we ring-fence the world of work from these challenging conversations then corporations can carry on with their singular focus on profit. No amount of "sustainability reporting" and "environmental audits" and "employee engagement exercises" can take away from the fact that adults who are denied a voice are not able to bring their best selves to resolving problems. Being an employee should not require us to sacrifice our agency. It is not enough to delegate corporate social responsibility to a specialist department that produces symbolic eco-packaging and commercials to associate the brand with tree-planting. This slows down the progress we need to make at changing the corporate mindset. Corporations can no longer choose how much or how little they engage with their local communities. Their employees are the ambassadors of the community. Their voices need to be heard – truly heard not as a part of a consultation exercise but as a part of everyday business.

Leaders who understand how to truly listen, how to respond, how to encourage participative problem-solving and how to share gains as well as pains – those leaders are the ones who are critical in creating the changes that are needed in the corporate world.

6.5 New stories are growing inside us

I would never have found my courage if it hadn't been for you

The Cowardly Lion. Wizard of Oz. L. Frank Baum

In the Twenty First Century, stories are still a key part of what it is to be human. In corporate life, in the political world and in societies across the globe, stories are used to enrol and engage large groups of strangers so that they work together. The stories we tell and the stories we follow can show us a way to escape our current peril or they can lead to the evolutionary scriptwriters eliminating us from the plot. We humans often forget we are only character actors: we have a bit part in a much longer story. We are at risk of not making it into the next episode.

Let's look at the plot line clichés of some familiar stories from our recent history:

The danger seems insurmountable. All hope is lost and then...

- The cavalry comes over the hill
- The worthless beans Jack picked up at the market turn out to be a magical route to riches.
- James Bond remembers the secret weapon that is just right for escaping from certain death: the wristwatch with rotating blades that cuts through the rope binding his wrists; the poison darts in his cigarettes, the ejector seat in his car
- "You're our only hope, Obi Wan Kenobi"
- Bilbo finds a ring that makes him invisible just as the orcs come down the tunnel.
- The invading Martians die out from the common cold

The Human Story is at that stage – maximum danger and no apparent way out. How is the cliché storyline going to save us from peril?

- A hero with a secret weapon?
- A special child with a scarred forehead and magical powers?
- Unexpected reinforcements?

Of course, deep down we know that no one is going to come from over the hill to rescue us from our perilous situation. There is no secret weapon or magical power that will wish all this threat away before bedtime. Wishful thinking will not rescue us. We need to sort this out for ourselves. Our story-telling capacity can help – but not by lazily waving a wand to wish all our problems away.

We need a different kind of magic to get us out of harm's way. Stories often have magical objects or magical powers or magical coincidences but there is no magic goblet that will wash away all the excess CO_2 we have released. There is no special ointment that will restore all the species we are driving to extinction. But we are powerful and we can make changes happen.

We humans need to make rapid changes together. Stories will be a key part of that massive alteration in our way of life. Stories are often repositories of wisdom: they have given us a menu inside our brains of how we respond to adversity.

Stories tell us we are all in it together. They can tell us where we are now. They can also show us how we got here and where we need to move to next. They set our expectations of how we escape from danger. It is not accidental that the last decade has seen a resurgence of gloomy novels and films set in post-apocalyptic futures. Young adult fiction has been dominated by dystopic tales. Our story world is starting to grapple with our predicament. But the stories so far have largely highlighted the predicament rather than pointed to a way through the danger to a sustainable future. Something isn't quite there yet in our imaginations but it is definitely forming, emerging, seeking expression. Soon these new stories will be informing our actions.

We Humans School Timetable
The School for Trying To Understand Human Beings

Choose one subject from
Column A and one from **Column B**

COLUMN A	COLUMN B
Social Anthropology	Leadership
Palaeo-Anthropology	Followership
Archaeology	Story telling
Human Biology	
Human Genetics and Evolution	
Sociology	
Economics	
Neuropsychology	
Social Psychology	
Evolutionary Psychology	
Demography	
Economics	
History	
Philosophy	

These change the world

These explain the world

6.6 We and We - not Us against Them

The moment of crisis has come.
We can no longer prevaricate.

David Attenborough, 16 January 2020

We humans can do some nasty things to each other when we act as a group. There is often a leader telling scary stories that brings the group together. When we hear that "They" are to blame then the people who were our neighbours become our enemies and the source of all our distress. This bonding around a story that labels another group as "the enemy" releases the worst in us. People with no previous disposition toward violence or hostility join in acts of terrible bloodshed. The story and the leader interfere with people's moral compass so they act in ways that would be unimaginable before they fell under the spell. This is not to justify the terrible sectarian slaughters which have scarred our history. It highlights the power of the bonds that bind co-believers in a story and a leader. The bonds bringing people together are strengthened when there is also a clearly identified enemy.

The "Leaders and Their Stories" glue is highly effective at binding groups to tackle adversity together. This is a behavioural adaptation to an environment where danger could best be tackled collectively. The dangers arising during our evolutionary history have left us with a tendency to unite into groups against perceived threats. Collaboration between different groups has been a recurring human pattern. Hostility between groups has also been an occasional part of the mix but we are not genetically determined to see other groups of our own species as threats who need to be defeated. Whatever our evolutionary history, it doesn't help us survive now. It is even more unhelpful if we are accustomed to seeing the natural environment as a threat. These are legacy patterns that we quickly need to modify. We are not fatalistically programmed to fight each other. We have a strong tendency toward leadership, followership and the use of stories to mediate our group relationships. But the positive aspect of our history is our flexibility. We have many and varied ways of living out our social needs: so many different groups, so many kinds of leadership.

Story-telling, leadership and followership remain our human superpowers. Without leadership and effective followership, without stories swiftly spreading new ideas, new loyalties and strong emotional bonds, we humans will not escape from this precarious situation.

It is good that we form strong group bonds. The challenge is to develop these bonds in a wider context. It is no longer helpful to create strong emotional ties as one group fighting against other groups. Nor can we come together to "do battle with nature".

We have to recognise it is no longer

"Us against Them".

It is

"We and We and We… We Humans, We Animals and Plants, We Elements of the Planet"

We are all in it together. We need leaders to bind us together with inclusive stories. We need to be brave in choosing which stories we follow through to the end.

The current stories our leaders are telling us may have tinges of green added for effect but these old-fashioned stories of battling against an imaginary enemy will not have happy endings. We need a radically different set of leadership behaviours and wonderful new stories showing us the way to living happily-ever-after.

Acknowledgements

I would like to thank Beth McCloskey for her
guidance and support during the writing of this book.

The following people have kindly read
earlier drafts and made helpful suggestions:

Christian Koenig, Roger Martin, Chris Breen and Ian Campbell.

Thanks to Ben, Dan, Jess, Niles, Clare and Phia.

Sam Hepburn is the wonderful cartoonist who has added
much-needed light relief to the difficult subject of extinction.

My brother John Clipson has helped in many practical ways.

I would like to express my gratitude to the following people
and institutions who have influenced the ideas in this book:

The Oxford Human Sciences Course.

Satish Kumar, Stephan Harding, Fritjof Capra and
the excellent work of Schumacher College.

The late AH "Chelly" Halsey.

My work in the corporate world has exposed me to good
"we are all in this together" leaders and poor "it's all about me"
leaders. I have learnt from the good and the bad. My thanks
to them all. I have learnt even more from my fellow followers.

About the Author

Paul Clipson is a citizen of the UK and the USA. In recent times he has watched both his countries fall under the spell of leaders peddling false and divisive stories.

Paul has enjoyed a thirty-year career in leadership development and organisational change. He knows how demoralizing it is to work for an appalling boss. His work with good bosses has given him hope.

He is a lifelong student of the Human Sciences, working across academic boundaries to find solutions to the challenges of our times.

Neander T Hal
@NeanderH

Homo sapiens – you are the future of humanity. The rest of us are gone.
It's not looking good, is it?
Unless you can learn to work together...
#WeHumans